Busy Ant Maths

Pupil Book 3A

Series Editor: Peter Clarke

Authors: Elizabeth Jurgensen, Jeanette Mumford, Sandra Roberts

Contents

2-digit numbers

Understand the place value of 2-digit numbers

1 Write the numbers shown by the Base 10.

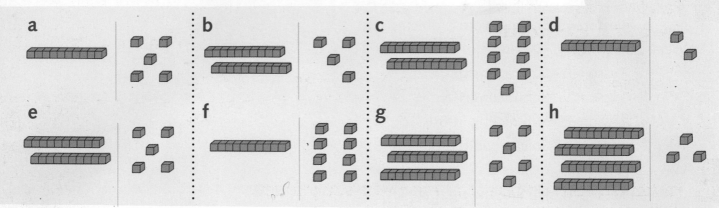

a   b   c   d

e   f   g   h

1 Write the numbers shown by the Base 10.

a   b   c   d

2 Write the numbers that have been separated into 10s and 1s.

a	50	4		b	30	9		c	7	40		d	9	70		e	2	20
f	80	3		g	50	0		h	3	60		i	50	9		j	8	90

1 Write these numbers in order, smallest first: 73, 13, 37, 33, 77, 17, 97, 7, 53

2 Draw this number line. Write the numbers from Question 1 on the number line.

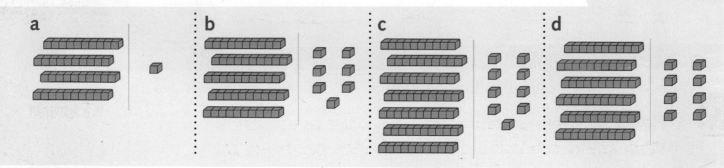

0   50   100

3 Write another number in between each number on the number line.

4 Explain the place value of 2-digit numbers. Imagine you are writing this for children in the year group below you.

Partitioning 2-digit numbers

Partition 2-digit numbers

Example

$20 + 6 = 26$

Partition these numbers into 10s and 1s using Base 10 material.

a   14   b   19   c   28   d   35   e   23

f   31   g   29   h   18   i   34   j   46

Partition these numbers in as many ways as you can. Write the calculation each time. If you need to, use Base 10 material to help you.

a   35   b   37   c   46   d   53   e   48

f   39   g   59   h   63   i   44   j   57

1   Partition these numbers in as many ways as you can. Write the calculation each time. Use the pattern of the calculations to help you.

a   57   b   68   c   71   d   88   e   95

Example

42

$40 + 2 = 42$

$30 + 12 = 42$

$20 + 22 = 42$

$10 + 32 = 42$

2   Choose a number from Question 1 and explain the pattern of the calculations.

3   Find the missing numbers.

a   ___ $+ 13 = 53$ b   ___ $+ 25 = 45$ c   ___ $+ 31 = 71$

d   ___ $+ 28 = 68$ e   ___ $+ 42 = 92$ f $30 +$ ___ $= 76$

g $50 +$ ___ $= 89$ h $60 +$ ___ $= 74$ i $20 +$ ___ $= 57$

5

3-digit numbers

Understand the place value of 3-digit numbers

Challenge 1

1 Write the numbers shown by the Base 10.

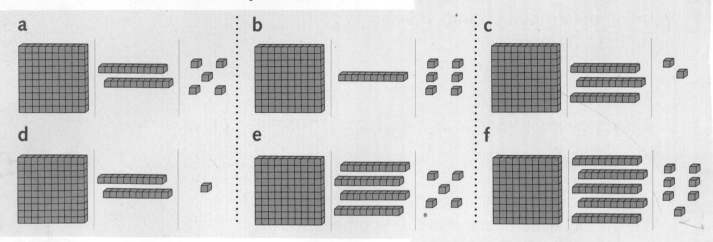

a

b

c

d

e

f

Challenge 2

1 Write the numbers shown by the Base 10.

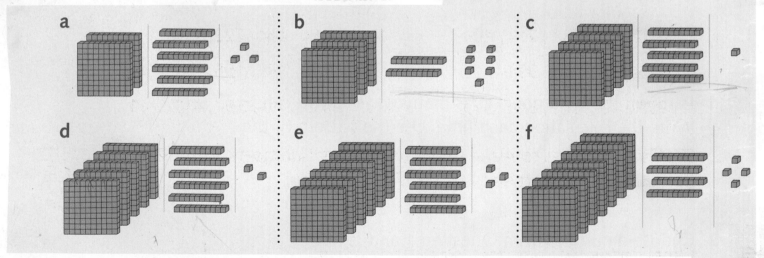

a

b

c

d

e

f

2 Write the numbers that have been separated into 100s, 10s and 1s.

   a 200 60 4   **b** 300 90 8   **c** 400 0 3   **d** 600 50 3

Challenge 3

1 Write the numbers that have been separated into 100s, 10s and 1s.

   a 7 700 40   **b** 20 1 900   **c** 5 500 70   **d** 3 60 100

2 I'm thinking of a number. The 100s digit is lower than 3, the 10s digit is odd and the 1s digit is 5. Write as many possible answers as you can.

Ordering numbers to 1000

Order numbers to 1000

Challenge 1

1 Order each set of numbers, smallest to largest.

a 54, 21, 87, 39

b 66, 25, 93, 12

c 74, 47, 17, 14, 41

d 105, 95, 107, 100, 101

2 These numbers are in order. What could the missing numbers be?

a ⬜ , 15, 38, ⬜ , 55

b 78, ⬜ , 84, ⬜ , ⬜

c 93, ⬜ , 100, ⬜ , 107

d 100, ⬜ , 112, 120, ⬜

Challenge 2

1 Order each set of numbers, smallest to largest.

a 142, 102, 241, 214, 204

b 214, 254, 223, 256

c 218, 327, 265, 376

d 362, 284, 155, 237

e 271, 207, 317, 301, 377

f 413, 314, 354, 412, 241

2 Use the digit cards to make six different 3-digit numbers.
Now put the numbers in order, smallest to largest.

Challenge 3

1 Order each set of numbers, smallest to largest.

a 634, 701, 792, 751, 603

b 701, 638, 739, 822, 811

c 912, 219, 903, 293, 921

d 875, 578, 998, 987, 919

2 a Use these digit cards to make all the possible 3-digit numbers.
Explain how you know you have all the possible numbers.

b Now put the numbers in order, smallest to largest.

Adding 2-digit numbers

Add mentally two 2-digit numbers

Challenge 1

1 Work out these calculations. Copy the number lines to help you.

a 23 + 21

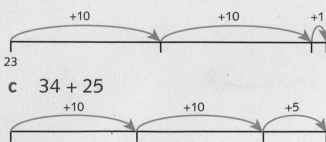

b 27 + 22

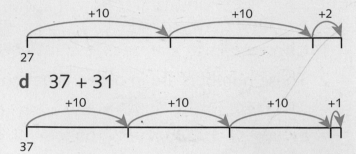

c 34 + 25

d 37 + 31

2 Now draw number lines for these calculations.

a 35 + 23   **b** 44 + 32   **c** 41 + 27   **d** 38 + 26

Challenge 2

1 Work out these calculations. Show your working out.

a 36 + 27   **b** 43 + 38   **c** 47 + 32   **d** 53 + 28

e 59 + 34   **f** 65 + 26   **g** 61 + 54   **h** 72 + 45

2 Can you work out these calculations mentally, without a number line?

a 89 + 43   **b** 85 + 57   **c** 75 + 45   **d** 93 + 38

Challenge 3

1 Work out these calculations. Show your working out.

a 74 + 58   **b** 83 + 65   **c** 89 + 76   **d** 93 + 66

e 68 + 83   **f** 49 + 78   **g** 94 + 57   **h** 83 + 85

2 Find the missing numbers.

a ☐ + 64 = 93   **b** ☐ + 46 + 87   **c** 53 + ☐ = 99

d 68 + ☐ = 115   **e** ☐ + 73 = 126   **f** ☐ + 62 = 134

Adding to 3-digit numbers

Add mentally 1s and 10s to a 3-digit number

Challenge 1

 40   3   26   50   23 5  44 32   10 41 4 8   30   7 20

1 Choose a 2-digit number and a 1s number from the circles. Write down the calculation. Do this ten times.

2 Choose a 2-digit number and a tens number from the circles. Write down the calculation. Do this ten times.

Examples

26 + 5 =

26 + 50 =

Challenge 2

 3   80   373   582 7   50 60   5   407   134 70   8 40   9 226

1 Choose a 3-digit number and a 1s number from the circles. Write down the calculation. Do this ten times.

2 Choose a 3-digit number and a 1s number from the circles. Write down the calculation. Do this ten times.

Examples

373 + 7 =

373 + 50 =

Challenge 3

 567   50   9   90   856 976   70 8   80 799   60   7   6   681   5

1 Choose a 3-digit number and a 1s number from the circles. Write down the calculation. Do this ten times.

2 Choose a 3-digit number and a 10s number from the circles. Write down the calculation. Do this ten times.

3 Write six different 3-digit add 10s calculations that equal 586.

Examples

681 + 8 =

681 + 70 =

Subtracting 2-digit numbers

Subtract mentally two 2-digit numbers

Challenge 1

1 Work out these calculations. Copy the number lines to help you.

a 35 – 13

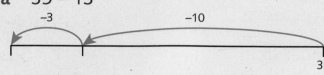

b 25 – 12

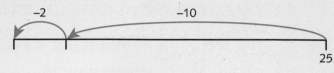

c 29 – 16

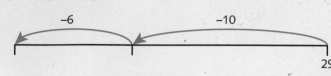

d 37 – 21

2 Now draw number lines for these calculations.

a 35 – 23   b 42 – 21   c 48 – 32   d 54 – 27

Challenge 2

1 Work out these calculations. Show your working out.

a 47 – 24   b 45 – 28   c 53 – 25   d 57 – 32

e 63 – 27   f 67 – 29   g 74 – 38   h 82 – 45

2 Can you work out these calculations mentally?

a 81 – 43   b 86 – 57   c 75 – 34   d 91 – 28

Challenge 3

1 Find the missing number in each calculation.

a 75 – ▢ = 32   b 83 – ▢ = 47   c 91 – ▢ = 55

d ▢ – 45 = 37   e ▢ – 28 = 63   f ▢ – 34 = 59

2 Explain how you worked these out.

Subtracting from 3-digit numbers

Subtract mentally 1s and 10s from 3-digit numbers

Challenge 1

28   20   7   46   51   10   30   40   2   3   6   59   32   4   50

1 Choose two numbers from the circles and write a 2-digit number subtract a 1s number calculation. Do this ten times.

2 Choose two numbers from the circles and write down a 2-digit number subtract a 10s number calculation. Do this ten times.

Examples

26 − 5 =
26 − 50 =

Challenge 2

70   5   9   252   7   60   507   50   314   275   8   80   40   486   3

1 Choose two numbers from the circles and write a 3-digit number subtract a 1s number calculation. Do this ten times.

2 Choose two numbers from the circles and write a 3-digit number subtract a 10s number calculation. Do this ten times.

Examples

373 − 7 =
373 − 50 =

Challenge 3

803   70   90   8   649   50   7   60   5   682   80   511   9   6   753

1 Choose two numbers from the circles and write a 3-digit number subtract a 1s number calculation. Do this ten times.

2 Choose two numbers from the circles and write a 3-digit number subtract a 10s number calculation. Do this ten times.

Examples

681 − 8 =
651 − 70 =

3 Write six different 3-digit subtract 10s calculations that equal 463.

Naming 3-D shapes

Recognise and name 3-D shapes lying in any position

Challenge 1

Match each picture to its 3-D shape. Copy and complete the table.

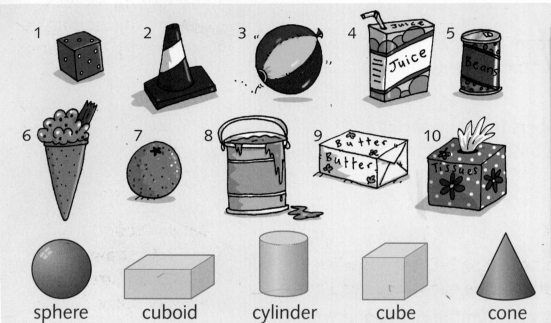

Picture	3-D shape
1	cube
2	
3	
4	
5	
6	
7	
8	
9	
10	

sphere   cuboid   cylinder   cube   cone

Challenge 2

1 Name the six 3-D shapes that are lying in the sand.

2 Use the clues to name the three shapes below.

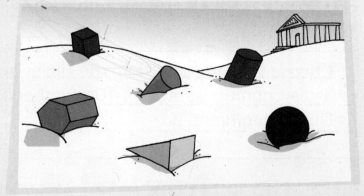

Shape A
- 6 rectangular faces
- 8 vertices
- 12 edges

Shape B
- 6 rectangular faces
- 2 triangular faces
- 9 edges

Shape C
- flat circular face
- curved sides that come to a point

Challenge 3

Write three clues to help identify each shape.

a   cube

b   hexagonal prism

c   square-based pyramid

d   cylinder

12

Making models of 3-D shapes

Make models of 3-D shapes using straws and 2-D shapes

Challenge 1

Work with a partner.

Make models of these 3-D shapes with the interlocking 2-D shapes.

  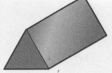 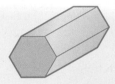

Challenge 2

Work with a partner.

Make skeletal models of these 3-D shapes with straws and sticky putty. Name each shape you make.

3-D shape	Long straws	Short straws	Sticky putty blobs
cube	12	0	8
a	8	4	8
b	3	6	6
c	3	3	4
d	5	10	10
e	4	4	5

Example

cube

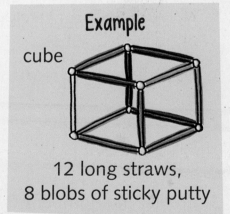

12 long straws,
8 blobs of sticky putty

Challenge 3

Work with a partner.

Using interlocking triangles and squares only:

    a explore the different 3-D shapes you can make

    b find a way to make a hexagonal prism.

Classifying and describing 3-D shapes

Sort and describe 3-D shapes

A cube

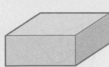

B cuboid

C cone

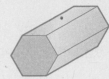

D hexagonal prism

E cylinder

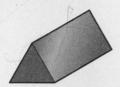

F triangular prism

G tetrahedron

H square-based pyramid

Challenge 1

Copy and complete the table.

3-D shape	A	B	C	D	E	F	G	H
Prism	✔							
Not a prism								

Challenge 2

1 Look at the faces of shapes **A** to **H** above.
   Write the names of shapes with **1 or more faces** that are:

   a square   **b** triangular   **c** rectangular   **d** curved

2 Copy and complete the table.

Prism end face	Number of sides of end face	Total number of edges
Triangle	3	9
Square		
Pentagon		
Hexagon		

3 Predict the number of edges for a prism with an end face of:

   a 8 sides   **b** 10 sides   **c** 12 sides

Challenge 3

Is there a relationship between the number of vertices and the number of edges of a prism? Investigate.

Building models with cubes

Build 3-D shapes with cubes

You will need:
• interlocking cubes

1 Build each model with 3 cubes. Count the number of square faces for each colour.

2 Copy and complete the table.

3-D model	Number of square faces		
	Red	Blue	Yellow
A	5		
B			
C			
D			

A   B   C   D

1 Build each model with 4 cubes. Count the number of square faces for each colour.

2 Copy and complete the table.

3-D model	Number of square faces			
	Red	Green	Blue	Yellow
A				
B				
C				
D				

A  B

C  D

1 Build these models with cubes. Continue until the 6th model.

2 Record your results in a table. Look for a pattern.

3 How many cubes will you need for the 10th model?

Counting in 2s, 3s, 5s and 10s

Count in multiples and steps of 2, 3, 5 and 10

Challenge 1

Copy and complete the table, writing in the multiples of each number.

2	4									
3			15							
5								50		
10					70					

Challenge 2

Help the frog. Write the numbers in the pattern. What is the last number?

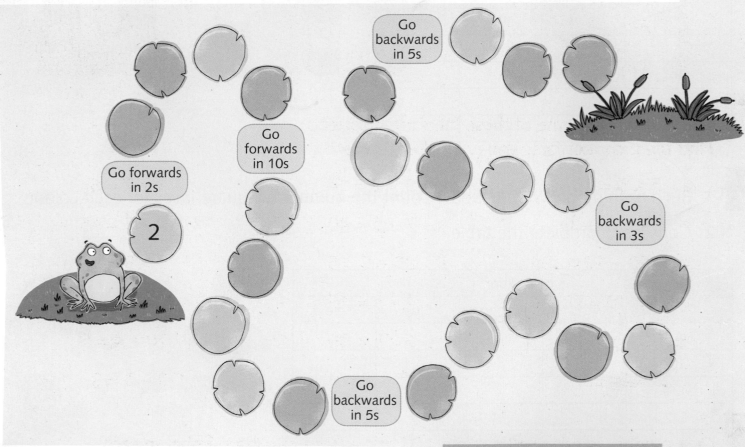

Challenge 3

Choose a start number from the grey box:

9   7
11   6   23

Choose a step number from these footprints. Count on from your start number in your chosen step. Continue the count for 10 numbers. Repeat for each start number. Make sure you use each step number at least once.

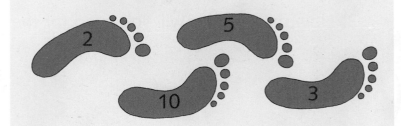

16

2s, 5s and 10s

Recall the multiplication and division facts for the 2, 5 and 10 multiplication tables

Challenge 1

Write a multiplication fact for each number coming out of the machine.

1
4
8
7
3
9

× 10
in 1
out→

2
6
9
4
7
8

× 2
in 1
out→

3
9
5
8
6
7

× 5
in 1
out→

Challenge 2

The answers to some of these facts are incorrect.
Find the incorrect facts and write them correctly.

a   5 × 5 = 25   b   8 × 2 = 14   c   3 × 2 = 6

d   24 ÷ 2 = 12   e   7 × 5 = 40   f   9 × 5 = 45

g   100 ÷ 10 = 10   h   30 ÷ 6 = 5   i   5 × 3 = 12

j   7 × 10 = 70   k   4 × 5 = 15   l   60 ÷ 5 = 11

m   40 ÷ 5 = 9   n   6 × 5 = 30   o   20 ÷ 4 = 5

p   9 × 3 = 37   q   25 ÷ 5 = 6   r   3 ÷ 3 = 3

Challenge 3

One number in each trio is missing. Work out the missing number in each set of trios, then write two multiplication and two division facts for each.

a
7
? 35

b
?
5 45

c
2
14 ?

d
5
? 60

e
8
80 ?

f
9
? 18

3 multiplication table

Recall the multiplication and division facts for
the 3 multiplication table

1 a $1 \times 2 =$ **2 a** $1 \times 5 =$ **3 a** $1 \times 10 =$ **4 a** $1 \times 3 =$

b $2 \times 2 =$ **b** $2 \times 5 =$ **b** $2 \times 10 =$ **b** $2 \times 3 =$

c $5 \times 2 =$ **c** $5 \times 5 =$ **c** $5 \times 10 =$ **c** $5 \times 3 =$

d $10 \times 2 =$ **d** $10 \times 5 =$ **d** $10 \times 10 =$ **d** $10 \times 3 =$

Challenge 2

Find the missing number in each calculation.

a ● $\times 3 = 12$ **b** $8 \times$ ⬡ $= 24$ **c** $6 \times$ ● $= 18$ **d** ▲ $\times 3 = 36$

e $11 \times$ ■ $= 33$ **f** $15 = 3 \times$ ▲ **g** ▲ $\times 3 = 21$ **h** $27 \div$ ■ $= 3$

i $9 \div$ ⬡ $= 3$ **j** ⬡ $\div 3 = 5$ **k** $9 =$ ■ $\times 3$ **l** $12 \div$ ● $= 3$

2 Write two multiplication and two division facts for each set of trios.

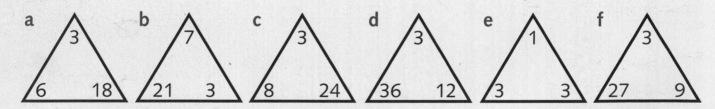

Challenge 3

Read the clues to find the number.

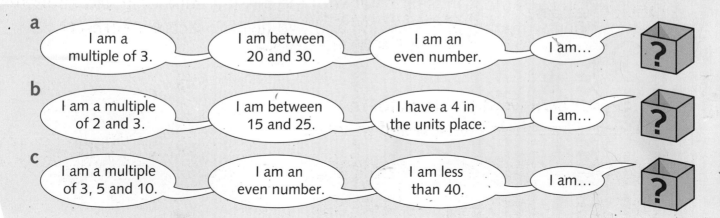

a I am a multiple of 3. | I am between 20 and 30. | I am an even number. | I am…

b I am a multiple of 2 and 3. | I am between 15 and 25. | I have a 4 in the units place. | I am…

c I am a multiple of 3, 5 and 10. | I am an even number. | I am less than 40. | I am…

Solving word problems (1)

Solve word problems and reason mathematically

Challenge 1

1 a 4 × 3 =

  b 40 × 3 =

2 a 3 × 3 =

  b 30 × 3 =

3 a 3 × 2 =

  b 30 × 2 =

4 a 5 × 3 =

  b 50 × 3 =

5 a 4 × 5 =

  b 40 × 5 =

6 a 6 × 5 =

  b 60 × 5 =

7 a 7 × 3 =

  b 70 × 3 =

8 a 9 × 3 =

  b 90 × 3 =

Challenge 2

1 Josh earns £3 pocket money per month. How much does he earn in 1 year?

2 There are 7 tricycles in the playground. How many wheels are there altogether?

3 Josh has 3 balloons. Mari has 6 times more balloons than Josh. How many balloons does Mari have?

4 Pizzas are split into 24 slices. Each child receives 3 slices. How many children are there?

5 9 children collect £30 each for the school charity. How much money is collected altogether?

6 In the cake sale, Year 3 collect £27 towards their class trip. They need to collect £12 more. What is the total amount they need to collect?

7 There are 21 bicycles in the playground. 3 are put into the shed. How many are left in the playground?

8 Mary and John eat a whole pizza. The pizza is split into slices. John eats 4 slices of pizza and Mary eats 5 slices of pizza. How many slices are there in the pizza?

9 12 cakes are sold on Saturday. 3 times more cakes are sold on Sunday. How many cakes are sold altogether at the weekend?

Challenge 3

Make up your own word problems for the calculations. Swap them with a friend to solve.

a 3 × 4 =

b 24 ÷ 3 =

c 50 × 3 =

d 27 ÷ 3 =

e 30 × 3 =

f 8 × 3 =

Finding fractions

Find fractions of a set of objects

1 Divide these groups into halves (2 equal groups).

a	8 children	**b**	4 apples
c	12 balls	**d**	10 shoes
e	14 oranges	**f**	20 pencils

2 Explain what $\frac{1}{2}$ means.

Example

$\frac{1}{2}$ of 6 = 3

1 Divide these groups into quarters (4 equal groups).

a	12 children	**b**	40 bananas
c	20 socks	**d**	16 bottles
e	24 lemons	**f**	4 cups

2 Explain what $\frac{1}{4}$ means.

Example

$\frac{1}{4}$ of 8 = 2

1 Divide these groups into thirds (3 equal groups).

a	12 leaves	**b**	6 plums	**c**	15 birds	**d**	9 books
e	18 children	**f**	24 cups	**g**	27 sweets	**h**	21 teddies

2 Divide these groups into fifths (five equal groups).

a	20 cups	**b**	25 oranges	**c**	15 books	**d**	30 pencils

3 Explain what the numerator and the denominator in a fraction tell us.

$+$ Numerator

$-$ Denominator

Solving fraction problems (1)

Solve a fraction problem

You will need:
- Resource 8: Square fractions (1)
- coloured pencils

Challenges 1,2

Use Resource 8: Square fractions (1).

1. How many ways can you find to shade half of the shape?

2. Check that all your ways of halving the shape are different. Cross out any that are the same.

3. Look at a partner's shapes. Have you both found the same number of ways of halving the shape?

I have found one way to shade half.

Challenge 3

Use Resource 9: Square fractions (2).

1. How many different ways can you find to shade half of the shape?

2. Check that all your ways of halving the shape are different.
Cross out any that are the same.

3. Draw in a systematic way all the different ways that you have found.

4. Explain how you know you have found all the shapes.

You will need:
- Resource 9: Square fractions (2)
- coloured pencils

21

Solving fraction problems (2)

Solve fraction problems and reason mathematically

Challenge 1

1 What fraction of these sweets is in each colour?

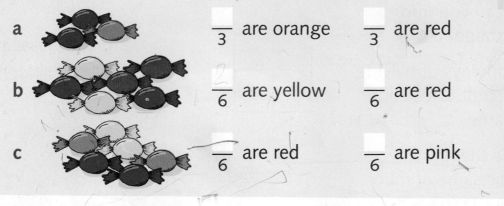

a $\frac{}{3}$ are orange $\frac{}{3}$ are red

b $\frac{}{6}$ are yellow $\frac{}{6}$ are red

c $\frac{}{6}$ are red $\frac{}{6}$ are pink $\frac{}{6}$ are yellow

2 Draw 5 sweets. Colour them in two different colours. Write the fraction for each colour of sweets. Remember that each sweet is $\frac{1}{5}$.

Challenge 2

1 What fraction of sweets is in each colour? Remember to count all the sweets first to find out the denominator.

a    b    c 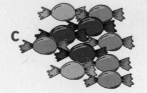   d

2 Draw 8 sweets. Colour them in. Write the fraction for each colour of sweets. Remember that each sweet is $\frac{1}{8}$.

3 Draw 5 more sweets and colour them differently. Write the fraction for each colour of sweets.

Challenge 3

1 Look at these sweets. What fraction of sweets is in each colour?

a    b

2 Look at picture **b**. What do you notice about the blue sweets?

3 In my sweet tube, $\frac{1}{2}$ are red and $\frac{1}{4}$ are yellow. How many sweets could I have?

Adding fractions

Add fractions with the same denominator

You will need:
- 5 interlocking cubes of 2 different colours

Challenge 1

1 Make a rod with 5 cubes and draw it in your book.

2 Write the fraction addition for your rod.

3 Make one more rod out of 5 cubes.
Draw it and write the fraction addition.

4 Make three rods using 6 cubes.
Draw them and write the fraction additions.

Challenge 2

1 Look at these rods. Write the fraction addition to go with each one.

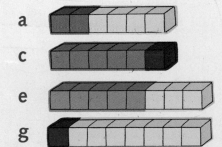

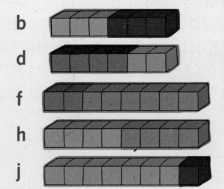

Example

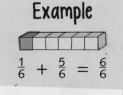

$\frac{1}{6} + \frac{5}{6} = \frac{6}{6}$

2 Draw three rods of your own and colour each one in 2 colours.
Write the fraction addition to go with each rod.

Challenge 3

1 a $\frac{1}{4} + \frac{3}{4} =$ b $\frac{2}{5} + \frac{3}{5} =$ c $\frac{1}{5} + \frac{4}{5} =$

  d $\frac{4}{6} + \frac{2}{6} =$ e $\frac{3}{6} + \frac{3}{6} =$ f $\frac{4}{7} + \frac{3}{7} =$

2 Explain why $\frac{7}{7}$ is the same as one whole.

3 a $\boxed{} + \frac{2}{3} = \frac{3}{3} =$ b $\frac{2}{5} + \boxed{} = \frac{5}{5}$ c $\frac{3}{8} + \boxed{} = \frac{8}{8}$

Fruit stall kilograms and grams

Know how many grams are equal to $\frac{1}{2}$, $\frac{1}{4}$ and $\frac{3}{4}$ of 1 kg

Challenge 1

1 Write each mass in grams.
  Remember: 1 kg = 1000 g.

Example

$1\frac{1}{4}$ kg = 1000 g + 250 g = 1250 g

a $\frac{1}{4}$ kg   b $\frac{1}{2}$ kg   c $\frac{3}{4}$ kg   d $1\frac{1}{2}$ kg   e $1\frac{3}{4}$ kg   f $2\frac{1}{2}$ kg

Challenge 2

1 Write each mass in grams.

Example

4 kg 300 g = 4000 g + 300 g = 4300 g

a 3 kg 500 g   b 4 kg 750 g   c 1 kg 900 g   d 6 kg 600 g

e $7\frac{1}{2}$ kg   f $8\frac{3}{4}$ kg   g $4\frac{1}{4}$ kg   h 6 kg 250 g

2 Each child buys 1000 g of fruit. They choose from packs of grapes, strawberries and cherries. Copy and complete the table.

Child	Grapes	Strawberries	Cherries
Beth	1	2	1
Colin	1	0	3
Diana	0	4	0
Ellen	0	0	4
Fred	0	2	0
Grace	0	0	0

Example

500 g + 200 g + 200 g = 900 g

Beth needs to buy 1 pack of cherries to make 1000 g.

Challenge 3

1 Harry buys 5 packs of fruit, which have a total mass of 1 kg 800 g.
  Work out how many packs of each fruit type Harry bought.

2 Could Harry buy 6 different packs with a total mass of 1 kg 800 g? Investigate.

Market stall scales

Read scales marked in kg and in g

Challenge 1

Write the mass shown on each scale.

a    b 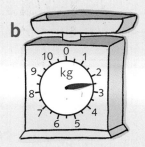   c    d

Challenge 2

1 Write the mass to the nearest 100 g, shown on each scale.

a    b 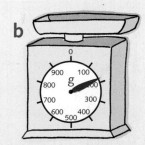   c 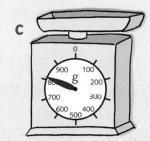   d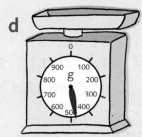

2 Write the mass shown on each scale in kilograms and then in grams.

a    b    c    d

Challenge 3

Fish A has a mass of 750 g. Use the clues to find the mass in grams of these fish.

a Fish B is 200 g lighter than Fish A.

b Fish C is double the mass of Fish B.

c Fish D is 350 g heavier than Fish C.

Puppy masses

Compare masses and multiples of mass in kg and g

Mass of puppies				
	At 3 months		At 6 months	
	kg	g	kg	g
Jack – Labrador	6	100	25	500
Mac – West Highland	2	300	6	900
Tack – Terrier	5	0	15	0
Zack – Spaniel	3	400	10	750

Challenge 1

1 Write the mass of each puppy at 3 months in grams.

  a Mac   **b** Tack   **c** Zack

Example

Jack

6 kg 100 g = 6000 g + 100 g = 6100 g

2 At 3 months, which puppy is:

  a The heaviest?   **b** The lightest?

  c 1100 g heavier than Mac?   **d** 2 kg 700 g heavier than Mac?

Challenge 2

1 At 6 months, which puppy had a mass of:

  a 6900 g?   **b** 15 kg?   **c** $25\frac{1}{2}$ kg?   **d** $10\frac{3}{4}$ kg?

2 Work out the mass that each dog has gained between 3 and 6 months.

Challenge 3

1 For how many days will a bag of Puppy Complete last if Jack is fed 500 g per day?

2 Find the total mass of puppy food when you buy 2, 5 or 10 packs of each item.

Puppy food	Buy 1	Buy 2	Buy 5	Buy 10
Puppy Complete	3 kg			
Chews	200 g			

Mass on the menu

Add and subtract mass in kg and g

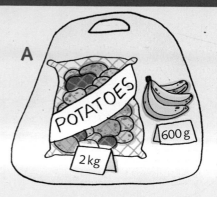

A

POTATOES

2 kg

600 g

B

BREAD

300 g

SOUP

800 g

C

750 g

400 g

1 Write the total mass of food in each bag in grams.

2 Find the difference in mass between shopping bags **B** and **C**.

Challenge 2

Look at the food in the shopping bags.

- Mrs Green bought potatoes, carrots and onions.
- Miss White bought two loaves of bread and three bananas.

1 How many grams heavier was Mrs Green's shopping than Miss White's?

2 Write the least number of weights you need to measure:

a 600 g of bananas   **b** 400 g carrots

c 750 g of onions   **d** 2 kg of potatoes

 50g   100g    200g   500g

Challenge 3

1 A trader has 6 boxes of goods. He needs to split them into 2 loads that are as equal in mass as possible. How might he do it?

8 kg   6 kg   7 kg   10 kg   5 kg   4 kg

2 A trader has 5 boxes with masses of 3 kg, 6 kg, 7 kg, 8 kg and 12 kg. How can he split them into:

a 2 equal delivery loads?   **b** 3 equal delivery loads?

Adding Is

- Add mentally a 3-digit number and 1s
- Solve missing number problems

Challenge 1

a $124 + 5 = \boxed{}$ b $157 + 4 = \boxed{}$ c $188 + 3 = \boxed{}$

d $169 + 8 = \boxed{}$ e $215 + 7 = \boxed{}$ f $303 + 9 = \boxed{}$

Challenge 2

1 Work out the missing numbers.

a $122 + \boxed{} = 127$ b $134 + \boxed{} = 138$ c $151 + \boxed{} = 158$

d $148 + \boxed{} = 156$ e $175 + \boxed{} = 181$ f $208 + \boxed{} = 215$

g $249 + \boxed{} = 258$ h $284 + \boxed{} = 295$ i $326 + \boxed{} = 335$

2 Explain your method for finding the missing numbers.

Challenge 3

1 Work out the missing numbers.

a $467 + \boxed{} = 472$ b $455 + \boxed{} = 463$ c $573 + \boxed{} = 582$

d $632 + \boxed{} = 640$ e $864 + \boxed{} = 873$ f $896 + \boxed{} = 903$

2 Solve these word problems. Write the calculation, work it out and then write the answer to the problem.

a Jack counts his favourite stickers. He has 238. His friend gives him some more and now he has 245. How many stickers does Jack's friend give him?

b Jamila has an apple tree in her garden. One day she picks 317 apples. She puts these with the apples she collected the day before. Altogether she has 326 apples. How many apples did she collect the day before?

c A sunflower grew to 278 cm in three weeks. One week later it is 286 cm tall. How much more has it grown?

Adding 10s

- Add mentally a 3-digit number and 10s
- Solve missing number problems

Challenge 1

a 124 + 30 = ☐　　**b** 157 + 20 = ☐　　**c** 195 + 40 = ☐

d 214 + 50 = ☐　　**e** 282 + 60 = ☐　　**f** 306 + 80 = ☐

Challenge 2

1 Work out the missing numbers.

a 118 + ☐ = 148　　**b** 137 + ☐ = 197　　**c** 185 + ☐ = 215

d 229 + ☐ = 289　　**e** 261 + ☐ = 321　　**f** 382 + ☐ = 432

g 467 + ☐ = 507　　**h** 498 + ☐ = 568　　**i** 611 + ☐ = 691

2 Explain your method for finding the missing numbers.

Challenge 3

1 Work out the missing numbers.

a 479 + ☐ = 529　　**b** 563 + ☐ = 613　　**c** 777 + ☐ = 847

d 865 + ☐ = 945　　**e** 819 + ☐ = 909　　**f** 944 + ☐ = 1004

2 Solve these problems. Write the calculation, work it out and then write the answer to the problem.

a Some children are hopping for charity. After 1 minute they have done 346 hops. Half a minute later they are up to 436 hops. How many hops did they do in the last half minute?

b In the snail race, one snail slithered 294 cm and the other snail slithered 344 cm. What was the difference between their distances?

3 Write four word problems for a partner.
Work them out yourself first so you know the answer!

Adding 100s

- Add mentally a 3-digit number and 100s
- Solve missing number problems

Challenge 1

1 a 24 + 200 = ____   **b** 76 + 100 = ____   **c** 53 + 200 = ____

d 69 + 300 = ____   **e** 81 + 500 = ____   **f** 93 + 400 = ____

2 a 152 + 100 = ____   **b** 164 + 100 = ____   **c** 186 + 100 = ____

d 169 + 200 = ____   **e** 157 + 200 = ____   **f** 204 + 200 = ____

g 395 + 300 = ____   **h** 412 + 300 = ____   **i** 475 + 100 = ____

Challenge 2

1 a 143 + 200 = ____   **b** 275 + 300 = ____   **c** 281 + 400 = ____

d 149 + 500 = ____   **e** 264 + 500 = ____   **f** 201 + 600 = ____

g 643 + 300 = ____   **h** 187 + 800 = ____   **i** 365 + 500 = ____

2 Work out the missing numbers.

a 127 + ____ = 427   **b** 182 + ____ = 482   **c** 265 + ____ = 665

d 315 + ____ = 715   **e** 458 + ____ = 858   **f** 544 + ____ = 844

g 795 + ____ = 995   **h** 106 + ____ = 906   **i** 639 + ____ = 939

3 Explain your method for finding the missing numbers.

Challenge 3

1 Work out the missing numbers.

a ____ + 500 = 931   **b** ____ + 200 = 603   **c** ____ + 400 = 536

d ____ + 700 = 812   **e** ____ + 900 = 927   **f** ____ + 600 = 1054

2 Explain your method for finding the missing numbers.

3 Write 10 missing number calculations for a partner. You need to know the answers so you can mark them!

Solving word problems (2)

Solve word problems and reason mathematically

Solve these word problems.

Challenge 1

a 74 people were on the train, 27 more got on.
How many people were on the train altogether?

b The ginger cat drank 63 ml of water.
The tabby cat drank 25 ml more. How much did the tabby cat drink?

c The shop sold 73 apples on Monday and 40 apples on Tuesday.
How many apples did they sell altogether?

d Two bananas cost 94p, three bananas cost 50p more.
How much do three bananas cost?

Challenge 2

a The chocolate cake costs 256p and the lemon cake costs 70p more.
How much does the lemon cake cost?

b Henry has 83 chocolates. He gives 27 away to friends.
How many does he have left?

c The bookshop sold 383 copies of the new adventure book in the morning
and 50 more copies in the afternoon. How many copies were sold that day?

d The train is very crowded. There are 425 passengers on it. 80 people get off
at the next stop. How many are on the train now?

e Louis' mum asks him to put his books away. He has 89 books.
He fits 43 on his bookshelf. How many does he still need to put away?

Challenge 3

a Henry has 120 chocolates. He gives 32 away to friends on Monday and 17
away on Tuesday. How many does he have left?

b Jamie and Julie have made 138 cards to sell at the school fair.
James made 59 of them. How many did Julie make?

c The teacher has 60 pencils. Each pencil pot must contain 5 pencils.
How many pencil pots can she fill?

d On Monday the sun shines for 187 minutes. On Tuesday it shines for
487 minutes. How many more minutes did the sun shine for on Tuesday?

Subtracting 1s

- Subtract mentally a 3-digit number and 1s
- Solve missing number problems

Challenge 1

a $127 - 3 =$ ☐   b $138 - 4 =$ ☐   c $131 - 5 =$ ☐

d $237 - 6 =$ ☐   e $285 - 7 =$ ☐   f $348 - 6 =$ ☐

Challenge 2

1 Work out the missing numbers.

a $129 - $ ☐ $= 125$ b $138 - $ ☐ $= 132$ c $169 - $ ☐ $= 161$

d $184 - $ ☐ $= 176$ e $276 - $ ☐ $= 270$ f $342 - $ ☐ $= 335$

g $529 - $ ☐ $= 522$ h $636 - $ ☐ $= 628$ i $741 - $ ☐ $= 732$

2 Explain your method for finding the missing numbers.

Challenge 3

1 Work out the missing numbers.

a $477 - $ ☐ $= 471$ b $591 - $ ☐ $= 582$ c $642 - $ ☐ $= 635$

d $763 - $ ☐ $= 755$ e $953 - $ ☐ $= 944$ f $802 - $ ☐ $= 796$

2 Solve these word problems. Write the calculation, work it out and then write the answer to the problem.

a Justin lives 235 metres from school. He has walked 227 metres so far. How much further does he need to go?

b Anne puts 435 ml of water in her cat's bowl. The cat has a drink. Now there is 429 ml left. How much water did the cat drink?

c Poppy collects stickers and has 530. She kindly gives some to her friend. Now she has 522. How many stickers did she give away?

Subtracting 10s

- Subtract mentally a 3-digit number and 10s
- Solve missing number problems

Challenge 1

a   148 – 30 = ☐   b   172 – 40 = ☐   c   184 – 50 = ☐

d   169 – 30 = ☐   e   197 – 60 = ☐   f   221 – 40 = ☐

Challenge 2

1 Work out the missing numbers.

a   186 – ☐ = 136   b   172 – ☐ = 112   c   195 – ☐ = 135

d   236 – ☐ = 186   e   371 – ☐ = 321   f   438 – ☐ = 388

g   484 – ☐ = 404   h   682 – ☐ = 592   i   649 – ☐ = 584

2 Explain your method for finding the missing numbers.

Challenge 3

1 Work out the missing numbers.

a   505 – ☐ = 455   b   685 – ☐ = 635   c   878 – ☐ = 798

d   901 – ☐ = 851   e   965 – ☐ = 885   f   1006 – ☐ = 986

2 Solve these word problems. Write the calculation, work it out and then write the answer to the problem.

a   432 children came into the hall for assembly. Some left for music club, 392 were left. How many children left for the music club?

b   Sam counts his money. He has £467. He pays for his football kit and then he has £397 left. How much did the kit cost?

3 Write 4 word problems for a partner.
Work them out yourself first so you know the answer!

Subtracting 100s

Subtract mentally a 3-digit number and 100s

Challenge 1

1 **a** 245 – 100 =   **b** 267 – 100 =   **c** 328 – 100 =

   d 461 – 100 =   **e** 395 – 100 =   **f** 421 – 100 =

2 **a** 267 – 200 =   **b** 328 – 200 =   **c** 362 – 200 =

   d 582 – 300 =   **e** 529 – 300 =   **f** 608 – 300 =

Challenge 2

1 **a** 345 – 200 =   **b** 497 – 200 =   **c** 506 – 300 =

   d 642 – 400 =   **e** 726 – 500 =   **f** 926 – 500 =

   g 947 – 700 =   **h** 852 – 800 =   **i** 843 – 600 =

2 Work out the missing numbers.

   a 658 –   = 258   **b** 593 –   = 193   **c** 649 –   = 349

   d 713 –   = 213   **e** 818 –   = 218   **f** 967 –   = 567

   g 682 –   = 182   **h** 957 –   = 57   **i** 403 –   = 103

3 Explain your method for finding the missing numbers.

Challenge 3

1 Work out the missing numbers.

   a   – 200 = 751   **b**   – 600 = 222   **c**   – 800 = 63

   d   – 700 = 195   **e**   – 300 = 666   **f**   – 500 = 1006

2 Explain your method for finding the missing numbers.

3 Write ten missing number calculations for a partner.
   You need to know the answers so you can mark them!

Solving word problems (3)

Solve problems and reason mathematically

Solve these word problems.

Challenge 1

a The green grocer had 73 bananas to sell.
The first customer bought a box of 40 bananas.
How many were left?

b Year 3 ate 89 apples in a week. Year 4 ate 30 fewer apples.
How many apples did Year 4 eat?

c The school ordered 245 pencils. Key Stage 1 took 100.
How many pencils were left for Key Stage 2?

d 135 passengers were on the train. After two stops, 50 people had got off.
How many people were left on the train?

Challenge 2

a The school has 642 pupils. Another 300 come to visit from a neighbouring
school. How many children were in school that day?

b Every day the cat is allowed 95 g of food. She is given 43 g for breakfast.
How much is left for her dinner?

c The whole school of 437 children go out on a visit to the seaside. On the
way home, one of the coaches breaks down and the 60 children on it are
late back. How many children got back on time?

d In the clothes shop, all jumpers cost £38. Jim buys a red jumper and a green
jumper. How much does he spend?

Challenge 3

a 485 children were already in assembly. Then 60 Year 3 children came
in late. How many were in the hall altogether?

b The banana cake weighs 642 g and the chocolate cake weighs 50 g less.
How much does chocolate cake weigh?

c There are 21 tomatoes. The teacher gives some children 3 each.
How many children were there?

d Red Class and Blue Class are having a competition to see who can collect
the most house points this term. Red Class has 237 so far, and Blue Class
has 637. How many more points does Blue Class have?

Finding the right angles

Find right angles in 2-D shapes

A 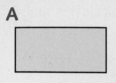   B   C 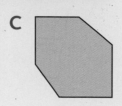   D 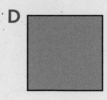   E 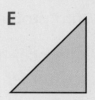   F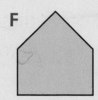

 Challenge 1

1 Use your right-angle tester to find the right angles in each shape.

2 Count the right angles.

3 Copy and complete the table.

Shape	A	B	C	D	E	F
Number of right angles	4					

You will need:
• right-angle tester

 Challenge 2

1 Make each shape on your pin board.

2 Use your right-angle tester and count the number of right angles.

3 Complete the table.

Shape	A	B	C	D	E	F
Number of right angles	4					

You will need:
• right-angle tester
• rubber band
• pin board

 Challenge 3

1 Draw these shapes on squared paper.

   a 3 different pentagons.

   b 3 different hexagons.

2 Circle the right angles in red pencil.

You will need:
• squared paper
• ruler
• red pencil

Turning patterns

Make and describe right-angled turns

You will need:
- squared paper
- ruler

Robby the robot can do three things:

- travel along straight lines

- turn clockwise through 1 right angle

- use a set of three numbers to make a pattern.

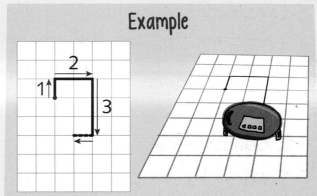

Example

This is a 1, 2, 3 pattern.

Challenge 1

1 Copy Robby the robot's 1, 2, 3 pattern onto squared paper.

2 Repeat the pattern three times.

3 Start each pattern where the last one finished.

Challenge 2

1 Divide the sheet of squared paper in half.

2 Mark a starting dot near the middle of each half.

3 Copy each of Robby's patterns and repeat it three times, starting where the last pattern finished.

    a The 1, 4, 2 pattern.   **b** The 2, 4, 1 pattern.

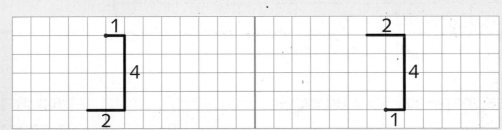

Challenge 3

1 Using the numbers 2, 3 and 5, make up three different patterns.

2 Draw the three patterns.

3 Write how the patterns are alike and how they are different.

Giving and following directions

Give and follow instructions to make turns

1 Write what you see when you have made these turns.

1 Face the garage.

   a $\frac{1}{4}$ turn to the right.   **b** $\frac{1}{2}$ turn to the right.   **c** $\frac{3}{4}$ turn to the right.

2 Face the duck pond.

   a $\frac{1}{4}$ turn to the right.   **b** $\frac{1}{2}$ turn to the right.   **c** $\frac{3}{4}$ turn to the right.

Copy and complete the tables.

1

I face the	I turn	I now face the
church	$\frac{1}{2}$ turn right	
shop	$\frac{1}{4}$ turn right	
duck pond	$\frac{3}{4}$ turn right	
garage	$\frac{3}{4}$ turn right	

2

I face the	I turn	I now face the
church		shop
shop		garage
duck pond		church
garage		duck pond

1 Write what you see after you have made these turns.

   a Face the shop. Make a $\frac{1}{2}$ turn to the right then a $\frac{1}{4}$ turn to the left.

   b Face the duck pond. Make a $\frac{1}{2}$ turn to the left then a $\frac{3}{4}$ turn to the right.

   c Face the garage. Make a $\frac{3}{4}$ turn to the right then a $\frac{1}{2}$ turn to the left and then a $\frac{1}{2}$ turn to the right.

2 Make up two questions that include making 2 or 3 turns.
Give them to a friend to work out.

All sorts of angles

Test whether an angle is greater than or less than a right angle

You will need:
- right-angle tester
- ruler

1 Test the size of these angles with your right-angle tester.

2 Copy and complete the table below. Write each letter in the correct column.

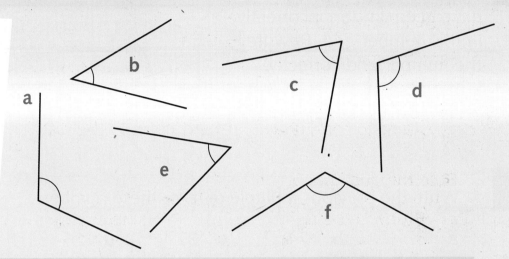

Less than a right angle	About a right angle	Greater than a right angle

1 Copy the table above.

2 Test the size of these kite angles with your right-angle tester.

3 Write each letter in the correct column.

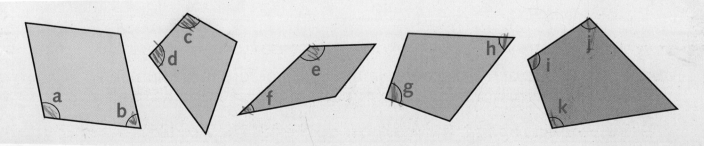

Look at the kite that has the angles k and j. Find a way to draw a similar kite where angles k and j are both right angles.

Use your right-angle tester to help you.

Counting in 4s

Count in multiples of 4

Challenge 1

Some of the multiples of 4 have been written incorrectly in this number grid. Rewrite the number grid correctly, putting the multiples of 4 in order, smallest to largest.

25	30	14	21
36	4	40	43
16	46	28	8

Challenge 2

1 Write the previous multiple of 4 for these numbers.

    a 8   **b** 24   **c** 32   **d** 16   **e** 20

2 Write the next multiple of 4 for these numbers.

    a 40   **b** 12   **c** 24   **d** 36   **e** 28

3 Write the multiple of 4 that is two multiples of 4 more than these numbers.

    a 4   **b** 20   **c** 32   **d** 8   **e** 28

4 Write the multiple of 4 that is two multiples of 4 less than these numbers.

    a 16   **b** 24   **c** 40   **d** 12   **e** 36

Challenge 3

Write the multiple of 4 that comes immediately before these numbers.

a 23   **c** 37   **e** 30   **g** 34

b 14   **d** 9   **f** 49

4 multiplication table

Recall the multiplication and division facts for
the 4 multiplication table

$1 \times 4 = 4$

$2 \times 4 = 8$

$5 \times 4 = 20$

$10 \times 4 = 40$

Challenge 1

1 Write down which of the key facts shown on the
keys you would use to answer the multiplication
facts below. Then write the answers to these facts.

a $9 \times 4 =$ **b** $4 \times 4 =$ **c** $3 \times 4 =$

d $6 \times 4 =$ **e** $8 \times 4 =$ **f** $7 \times 4 =$

2 Write the multiples of 4 from 4 to 48.

4, ___, ___, ___, ___, ___, ___, ___, ___, ___, ___, 48

Challenge 2

1 Write two multiplication and two division facts for each set of trios.

a 4 / 6 / 24   **b** 4 / 7 / 28   **c** 1 / 4 / 4   **d** 4 / 4 / 16   **e** 9 / 4 / 36   **f** 8 / 4 / 32

2 Find the missing number in each calculation.

a ● $\times 4 = 12$ **b** ▲ $\times 4 = 36$ **c** ■ $\times 4 = 28$

d ■ $\div 4 = 5$ **e** ⬡ $\div 4 = 10$ **f** $6 \times$ ⬡ $= 24$

g $11 \times$ ⬡ $= 44$ **h** $48 \div$ ■ $= 4$ **i** $36 =$ ▲ $\times 4$

Challenge 3

Read the clues to find the number.

1 I am a multiple of 3 and 4.   I am between 10 and 25.   I have a 4 in the units place.   I am...   ?

2 I am a multiple of 4, 5 and 10.   I am an even number.   I am less than 50 but greater than 20.   I am...   ?

Doubling to find the 4 multiplication table

- Use doubling to recall the 4 multiplication table
- Multiply a tens number by 4

Challenge 1

Double these numbers.

1 a 4   b 9   c 7   d 6   e 8   f 3   g 5

2 a 20   b 50   c 60   d 30   e 70   f 90   g 40

Challenge 2

1 Copy and complete the number line for the 2 multiplication facts.

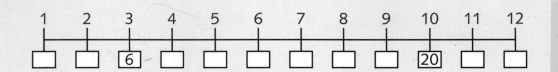

2 Double the answers to the 2 multiplication facts to work out the answers to the 4 multiplication facts.

Copy and complete the number line.

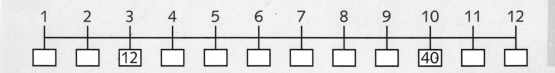

Hint

Use your completed number lines to recite the 2 and 4 multiplication and division facts, e.g.

$3 \times 2 = 6$,
$20 \div 2 = 10$,
$3 \times 4 = 12$,
$40 \div 4 = 10$.

Challenge 3

Write the multiplication fact for each number coming out of the machine.

1 a 40
   b 60
   c 90
   d 30
   e 70

2 a 40
   b 60
   c 90
   d 30
   e 70

Multiplication and division

Write a multiplication statement that matches a division statement

 Challenge 1

Find the missing number in each calculation.

$4 \times 9 = 36$

a ⭐ × 4 = 20

b 7 × ⭐ = 21

c 36 = ⭐ × 4

d 4 × 7 = ⭐

e 5 × ⭐ = 25

f ⭐ × 4 = 32

g 8 × ⭐ = 24

h 45 = 9 × ⭐

i 8 × ⭐ = 80

j 6 × 4 = ⭐

k ⭐ × 3 = 15

l 16 = ⭐ × 4

Challenge 2

1 Write a division fact for each multiplication fact in Challenge 1.

2 One number in each trio is missing. Work out the missing number in each set of trios, then write two multiplication and two division facts for each.

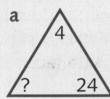

a
```
    4
  ?   24
```

b
```
    ?
 48    4
```

c
```
    2
 16    ?
```

d
```
    3
  ?   27
```

e
```
    ?
  7   28
```

f
```
    6
  ?   18
```

Challenge 3

Write the matching multiplication fact or division fact for each puzzle.

Example

5 × 6 = 30   30 ÷ 5 =

a 7 × 4 =

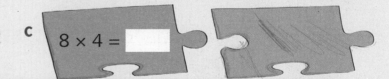

b 70 ÷ 7 =

c 8 × 4 =

d 36 ÷ 4 =

e 16 ÷ 4 =

Counting in 8s

Count in multiples of 8

Challenge 1

Some of the multiples of 8 are missing from the number grid. Rewrite the number grid correctly, putting the multiples of 8 in order, smallest to largest.

Challenge 2

1 Write the previous multiple of 8 for these numbers.

    a 16   **b** 48   **c** 64   **d** 80   **e** 24

2 Write the next multiple of 8 for these numbers.

    a 40   **b** 8   **c** 32   **d** 72   **e** 56

3 Write the multiple of 8 that is two multiples of 8 more than these numbers.

    a 8   **b** 32   **c** 72   **d** 48   **e** 64

4 Write the multiple of 8 that is two multiples of 8 less than these numbers.

    a 24   **b** 80   **c** 40   **d** 56   **e** 96

Challenge 3

Write the multiple of 8 that comes before these numbers.

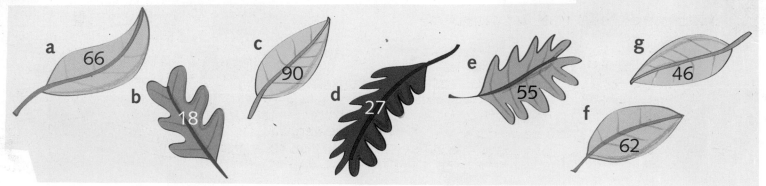

a 66   **b** 18   **c** 90   **d** 27   **e** 55   **f** 62   **g** 46

8 multiplication table

Recall the multiplication and division facts for
the 8 multiplication table

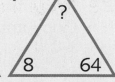

1 × 8 = 8

2 × 8 = 16

5 × 8 = 40

10 × 8 = 80

1 Write down which of the key facts you would
use to answer the multiplication facts below.
Then write the answers to these facts.

a 9 × 8 =   b 7 × 8 =   c 4 × 8 =

d 6 × 8 =   e 3 × 8 =   f 8 × 8 =

2 Write the multiples of 8 from 8 to 96.

8, [ ] , [ ] , [ ] , [ ] , [ ] , [ ] , [ ] , [ ] , [ ] , 96

1 One number in each trio is missing. Work out the missing number in each
set of trios, then write two multiplication and two division facts for each.

a ? / 8 24
b 8 / ? 16
c 7 / 8 ?
d 9 / ? 8
e 8 / 32 ?
f ? / 8 64

2 Find the missing
number in each
calculation.

a ★ × 8 = 40   b 8 × ★ = 24   c 8 × ★ = 72

d ★ × 8 = 32   e ★ × 8 = 64   f 48 ÷ ★ = 8

g 56 ÷ ★ = 8   h ★ ÷ 8 = 11   i 64 ÷ ★ = 8

Read the clues to find the numbers.

1 I am a multiple
of 3, 4 and 8.   I am less
than 40.   I am an
even number   I am...   ?

2 We are multiples
of 4 and 8.   We are greater than
20 but less than 40.   We are...   ?

Doubling to find the 8 multiplication table

- Use doubling to recall the 8 multiplication table
- Multiply a tens number by 8

Challenge 1

Double these numbers, then double them again.

1 a 6   b 3   c 9   d 5   e 7   f 4   g 8

2 a 50   b 20   c 80   d 60   e 70   f 40   g 90

Challenge 2

1 Copy and complete the number line for the 4 multiplication facts.

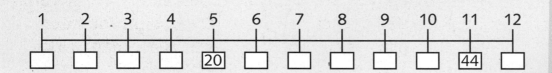

2 Double the answers to the 4 multiplication facts to work out the answers to the 8 multiplication facts.

Copy and complete the number line.

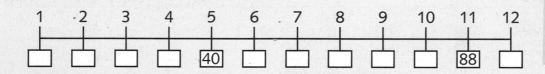

Hint

Use your completed number lines to recite the 4 and 8 multiplication and division facts, e.g.

$5 \times 4 = 20$,
$44 \div 4 = 11$,
$5 \times 8 = 40$,
$88 \div 8 = 11$.

Challenge 3

Write the multiplication fact for each number coming out of the machine.

1 a 30
  b 60
  c 70
  e 40
  f 90

2 a 30
  b 60
  c 70
  d 40
  e 90

Solving word problems (4)

Solve word problems and reason mathematically

Challenge 1

Find the missing number in each calculation.

a ● ÷ 4 = 5

b 7 × ▲ = 28

c 16 = ■ × 4

d 64 ÷ ■ = 8

e 5 × ⬡ = 40

f ⬡ ÷ 4 = 12

g 8 × ⬡ = 24

h 40 = 8 × ■

i 36 ÷ 4 = ▲

Challenge 2

Year 3 are having a bake sale to raise money for their favourite charity.

1 10 bags of popcorn cost £4. How much do 40 bags cost?

2 There are 32 biscuits on each tray. There are 8 biscuits in each row. How many rows of biscuits?

3 There are 24 chocolate cakes to sell. 8 children buy an equal number each. How many cakes do they buy?

4 There are 7 large cakes for sale. Each cake can be split into 8 slices. How many slices are there altogether?

5 9 children buy 4 cupcakes each. How many cupcakes are there altogether?

6 Cupcakes cost 40p each. Jana buys 5 and Rani buys 3. How much do they spend altogether on cakes?

7 The cake stall made 4 times more money than the popcorn stall. If the cake stall made £48, how much did the popcorn stall make?

8 Year 4 children spent £16 at the bake sale. They spent £12 less than Year 3 children. How much did Year 3 spend?

9 Year 3 made £96. They paid £8 back to the office for supplies and the rest went to charity. How much did they give to their charity?

Challenge 3

Make up your own word problems for the calculations. Swap them with a friend to solve.

a 8 × 4 =

b 60 × 4 =

c 40 × 4 =

d 36 ÷ 4 =

e 48 ÷ 8 =

f 64 ÷ 8 =

Up to the minute

Tell and write the time to the minute on a 12-hour clock with hands

  Write each time in words.

a    **b**    **c**    **d**

1 Use Resource 23 to show these times.

You will need:
- Resource 23: Blank clock faces Set A

a 9 minutes past 8   **b** 11 minutes to 4

c 22 minutes to 3   **d** 18 minutes past 11

2 Write each time in two ways.

a    **b**    **c**

Example

37 minutes past 3, 23 minutes to 4

Look at the clocks in Question 2 of Challenge 2.

1 Write what time each clock showed:

a 10 minutes ago   **b** 20 minutes ago   **c** 2 hours ago

2 Write what time each clock will show:

a in 15 minutes   **b** in 25 minutes   **c** in 2 hours

Using a time line

Use a time line and the vocabulary of time

1 Look at Joe's time line.

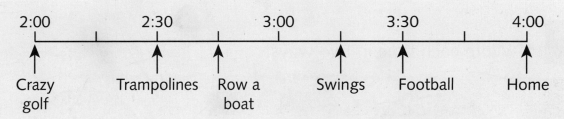

| 2005 | 2010 | 2015 |

Was born   Started school   Started Y3

Joe was born in _____. He started school in _____ and began Y3 in _____.

2 Draw a time line similar to Joe's about you.

Look at this time line of a day at the park.

| 2:00 | 2:30 | 3:00 | 3:30 | 4:00 |

Crazy golf   Trampolines   Row a boat   Swings   Football   Home

1 Write the time when the friends went to:

   a the trampolines   **b** the swings   **c** row a boat

2 Write what the friends were doing:

   a at 2:15   **b** at 3 o'clock   **c** at quarter to 4

3 About how long did the friends spend playing:

   a crazy golf?   **b** football?

Work with a partner.

Plan some activities you would like to do on a Saturday afternoon. Draw a time line from 2 o'clock onwards. Show with arrows and labels when you will do each activity.

49

Roman numerals and 24-hour times

Read the time to the minute on a clock with
Roman numerals and on a 24-hour clock

You will need:
- Resource 23: Blank clock faces Set A

Challenge 1

Use Resource Resource 23: Blank clock faces
Set A to show these times.

a 20 minutes past 8   **b** 12 minutes past 10   **c** 13 minutes to 4

d 28 minutes to 1   **e** 7 minutes past 9   **f** 8 minutes to 5

Challenge 2

1 Write each time in two ways.

a   **b**   **c**

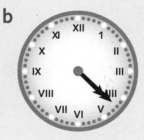

Example

41 minutes past 5,
24 minutes to 6

2 Write the 12-hour and 24-hour times for each clock.

a   **b**   **c**

Example

7:25
19:25

Challenge 3

1 Read the clues to find which time from the grey
box the clock shows. The time on the clock is:

- in the morning
- before 9 am
- after quarter past 8

9:10 am
8:28 am
8:46 pm
8:13 am

2 Write a clock puzzle similar to this. Ask a friend to solve it.

Timing tasks

Estimate and measure time in minutes

You will need:
- 20 plastic beads
- thread or lace
- 5-minute timer or stopwatch
- sheet of paper

Challenge 1

1 Estimate how many minutes it takes you to:

   a brush your teeth   **b** eat a sandwich   **c** tie the laces on your trainers

2 Work with a partner to complete Tasks 1 and 2.

Task 1: Write your name 20 times.

Task 2: Thread 20 plastic beads.

Take turns to:
- estimate how many minutes the task will take
- complete the task, with your partner timing it.

Challenge 2

1 Estimate how many minutes it takes you to:

   a eat a sandwich   **b** play a computer game

You will need:
- 25 pegs
- pegboard
- 5-minute timer or stopwatch

2 Work with a partner to complete Tasks 1 to 3.

Take turns to:
- estimate how many minutes the task will take
- complete the task, with your partner timing it.

Task 1: Count backwards in steps of 3 from 36 to 0.

Task 2: Place 25 pegs in a 5 × 5 square on a pegboard.

Task 3: Hop on one foot 10 times without stopping. If your other foot touches the floor you must start again but the timer keeps on running.

Challenge 3

1 Work with a partner. Estimate then time how many minutes it takes to build:

You will need:
- interlocking cubes
- timer or stopwatch

   a 4 cubes **b** 9 cubes **c** 16 cubes **d** 25 cubes

2 Look at how you organised the task in Question 1. Can you think of a way to complete it more quickly? Test your idea.

Maths facts

Problem solving

The seven steps to solving word problems
1 Read the problem carefully.   2 What do you have to find?
3 What facts are given?   4 Which of the facts do you need?   5 Make a plan.
6 Carry out your plan to obtain your answer.   7 Check your answer.

Number and place value

100	200	300	400	500	600	700	800	900
10	20	30	40	50	60	70	80	90
1	2	3	4	5	6	7	8	9

Addition and subtraction

Number facts

+	0	1	2	3	4	5	6	7	8	9	10
0	0	1	2	3	4	5	6	7	8	9	10
1	1	2	3	4	5	6	7	8	9	10	11
2	2	3	4	5	6	7	8	9	10	11	12
3	3	4	5	6	7	8	9	10	11	12	13
4	4	5	6	7	8	9	10	11	12	13	14
5	5	6	7	8	9	10	11	12	13	14	15
6	6	7	8	9	10	11	12	13	14	15	16
7	7	8	9	10	11	12	13	14	15	16	17
8	8	9	10	11	12	13	14	15	16	17	18
9	9	10	11	12	13	14	15	16	17	18	19
10	10	11	12	13	14	15	16	17	18	19	20

+	11	12	13	14	15	16	17	18	19	20
0	11	12	13	14	15	16	17	18	19	20
1	12	13	14	15	16	17	18	19	20	
2	13	14	15	16	17	18	19	20		
3	14	15	16	17	18	19	20			
4	15	16	17	18	19	20				
5	16	17	18	19	20					
6	17	18	19	20						
7	18	19	20							
8	19	20								
9	20									

Number facts

+	0	10	20	30	40	50	60	70	80	90	100
0	0	10	20	30	40	50	60	70	80	90	100
10	10	20	30	40	50	60	70	80	90	100	110
20	20	30	40	50	60	70	80	90	100	110	120
30	30	40	50	60	70	80	90	100	110	120	130
40	40	50	60	70	80	90	100	110	120	130	140
50	50	60	70	80	90	100	110	120	130	140	150
60	60	70	80	90	100	110	120	130	140	150	160
70	70	80	90	100	110	120	130	140	150	160	170
80	80	90	100	110	120	130	140	150	160	170	180
90	90	100	110	120	130	140	150	160	170	180	190
100	100	110	120	130	140	150	160	170	180	190	200

+	110	120	130	140	150	160	170	180	190	200
0	110	120	130	140	150	160	170	180	190	200
10	120	130	140	150	160	170	180	190	200	210
20	130	140	150	160	170	180	190	200	210	220
30	140	150	160	170	180	190	200	210	220	230
40	150	160	170	180	190	200	210	220	230	240
50	160	170	180	190	200	210	220	230	240	250
60	170	180	190	200	210	220	230	240	250	260
70	180	190	200	210	220	230	240	250	260	270
80	190	200	210	220	230	240	250	260	270	280
90	200	210	220	230	240	250	260	270	280	290
100	210	220	230	240	250	260	270	280	290	300

Written methods – addition

Example: 548 + 387

Expanded written method

```
   548
 + 387
    15
   120
   800
   935
```

Formal written method

```
   548
 + 387
   935
   1 1
```

Written methods – subtraction

Example: 582 – 237

Formal written method

```
     7 12
   5 8 2
 - 2 3 7
   3 4 5
```

Multiplication and division

Number facts

x	2	3	4	5	8	10
1	2	3	4	5	8	10
2	4	6	8	10	16	20
3	6	9	12	15	24	30
4	8	12	16	20	32	40
5	10	15	20	25	40	50
6	12	18	24	30	48	60
7	14	21	28	35	56	70
8	16	24	32	40	64	80
9	18	27	36	45	72	90
10	20	30	40	50	80	100
11	22	33	44	55	88	110
12	24	36	48	60	96	120

Written methods – multiplication

Example: 63×8

Partitioning

$$63 \times 8 = (60 \times 8) + (3 \times 8)$$
$$= 480 + 24$$
$$= 504$$

Grid method

×	60	3	
8	480	24	= 504

Expanded written method

```
    6 3
×     8
    2 4   ( 3 × 8)
  4 8 0   (60 × 8)
  5 0 4
  1
```

Formal written method

```
    6 3
×   ₂8
  5 0 4
```

Written methods – division

Example: $92 \div 4$

Partitioning

$$92 \div 4 = (80 \div 4) + (12 \div 4)$$
$$= 20 + 3$$
$$= 23$$

Expanded written method

```
      2 3
  4 ) 9 2
      8 0   20 × 4
      1 2
      1 2   3 × 4
        0
```

Formal written method

```
      2 3
  4 ) 9 ¹2
```

Fractions

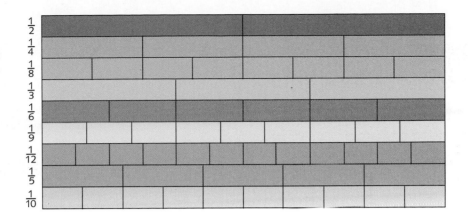

Measurement

Length
1 metre (m) = 100 centimetres (cm) = 1000 millimetres (mm)

Mass
1 kilogram (kg) = 1000 grams (g)

Capacity
1 litre (*l*) = 1000 millilitres (ml)

Time
1 year	=	12 months
	=	356 days
	=	366 days (leap year)
1 week	=	7 days
1 day	=	24 hours
1 hour	=	60 minutes
1 minute	=	60 seconds

12-hour clock

24-hour clock

Properties of shape

right-angled triangle   equilateral triangle   isosceles triangle   scalene triangle

circle   semi-circle   pentagon   hexagon   heptagon   octagon   square   rectangle

 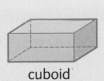

cube   cuboid   cone   cylinder   sphere   triangular prism   triangular-based pyramid (tetrahedron)   square-based pyramid